I0763039

THE

COPPER MINES

OF

LAKE SUPERIOR.

HYDE PARK:
PRINTED AT THE NORFOLK COUNTY GAZETTE OFFICE.
1873.

EXPLANATORY.

Office, No. 4 Congress Square
BOSTON, March, 1873.

This document has been prepared by the writer to satisfy, in a measure, the constant inquiry made in regard to the Copper Mines of Lake Superior, many of which figure in the daily transactions of the Boston Stock Exchange. The statistics have been compiled from good authorities, and the detailed statements of most of the mines have been derived from official sources.

I am gratified to be able to add to the pamphlet the written testimonials of prominent men connected with the Copper mining interests, and of firms which have long been dealers in most of the mines named, in favor of the general accuracy of the statements in the document.

If the information serves to awaken any interest in Copper mining, or to convey information in regard to its present condition and prospects, the mission of this pamphlet will have been accomplished.

O. D. ASHLEY.

THE COPPER MINES

OF

LAKE SUPERIOR.

COPPER mining on Lake Superior commenced in 1845, about twenty-eight years since. Prior to this date copper had been obtained throughout the world in the form of sulphuret. The discoveries of Lake Superior were of native copper, which was a novelty in copper mining, and so improbable, according to all geological precedents, that much doubt was expressed by scientific men in regard to its reality. The facts were, however, abundantly proven. The distribution of this native copper, which seemed such a geological wonder, was principally in masses of various weight; barrel work, which is strung together in adhering rock, like roots of a tree in the soil, and "stamp stuff," which is disseminated through the copper bearing rock in small particles, known among miners as "shot copper." In these three forms the native copper of Lake Superior is distributed.

The first mine from which profitable results were obtained was the once famous Cliff Mine, then worked by the Pittsburgh and Boston Mining Company, holding its office at Pittsburgh, Penn. During several years this mine paid large dividends, and has actually returned the sum of $2,280,000, upon an expenditure by the stockholders in assessments of

$110,000. Following this came the Minesota, which paid dividends amounting to $1,750,000, upon a total expenditure by the stockholders of $436,000. The product of these mines was principally mass copper and barrel work—but little account being made of the stamp stuff, which was wrought at that time under slow and expensive processes. The masses produced by these mines were of extraordinary size, and the respective veins were so rich that great excitement was created. The largest mass was found in the Minesota, and weighed more than five hundred tons, producing about ninety per cent. of ingot copper, but many masses weighing from fifty to two hundred tons were found in both mines.

These great successes, after some years of persistent mining and the expenditure of large sums of money, stimulated mining enterprise in the Lake Superior Copper District to such an extent that locations contiguous to either and supposed to be carrying the same veins were eagerly taken up, and the shares bought by the public, in the reasonable hope of similar results. The National and Rockland on either side of the Minesota, and the North American adjoining the Cliff, sold at very high prices, and a crowd of adventurers and speculators, taking advantage of the excitement, located upon more distant lands upon the same mineral range, in some instances having actually very good surface indications, and in others but little more than a mere location. These new mines were floated upon the Boston Stock Market, and many of them were honestly worked by the managers, who shared in the general feeling of confidence created by the profitable results from the Cliff and Minesota. Others were equipped and partially opened, with but little real basis for the expenditures made.

The table of assessments annexed to this pamphlet, taken from the Portage Lake Mining Gazette, will give an idea of the extravagant sums paid out upon mines which have not yielded a dollar in dividends, although in some cases a considerable amount of copper was taken out. Among them are the Albany and Boston, which expended $615,000; the Amygdaloid, $470,000; Bay State, $385,000; Bohemian, $343,-

000; Flint Steel River, $244,000; Hancock, $570,000; Huron, $500,000; Isle Royale, $910,000; North Western, $227,000; Norwich, $220,000; Pennsylvania, $500,000; Phœnix, $820,000; Rockland, $340,000; Sheldon, $400,000; South Pewabic, $500,000; Star, $225,000; Superior, $220,000; Toltec, $420,000. Such was the confidence in final results, and especially in such mines as the Hancock, Huron, Rockland, Sheldon, Isle Royale, Phœnix and Toltec, which produced encouragingly at the start, that money was called for and willingly paid to develop them, and while the managers, living in cities a thousand miles distant, where the money was furnished and the stocks bought and sold, had great faith in ultimate profitable results, and were perfectly honest in their efforts to secure them, they had not sufficient experience at the time to form trustworthy opinions, and were perhaps too sanguine in their expectations. The great bulk, however, of all the money expended in copper mining on Lake Superior, so far as those who directed the Companies controlled it, was honestly disbursed beyond question. There was great extravagance in machinery and surface improvements, upon the too positive assumption that brilliant returns would follow, and warrant the expense. The general impulse was to push development, the sooner to arrive at a dividend basis, and the mining agents urged by the directors expended money with lavish hands.

At this period the difficulties which mining companies were obliged to encounter were very serious,—almost enough in fact to discourage mining enterprise entirely. Supplies were obtained with difficulty and at great expense from Detroit, while copper had to be carried seven hundred miles before it reached the smelting works.

After some years of prosperity, the great Cliff and Minesota mines, having been opened to a depth which made it unprofitable to work them at the price then ruling for ingot copper, (17 cents,) were obliged finally to cease operations.

This caused great astonishment and great disappointment not only to the holders of these stocks, but to the owners of the numerous ventures which had started into life upon the prestige and success of these two leading mines. The usual consequences of an over-sanguine and wide-spread speculation followed, and one after another the copper mines of Lake Superior declined to skeleton figures in the stock market, or faded out of sight. The blow to copper mining on Lake Superior was very severe, and it lingered for some years in a flickering state, only sustained by the tribute system of mining, until very satisfactory results began to be shown on Portage Lake, especially in the Quincy, Pewabic and Franklin mines. Following this came the discovery and working of the famous Calumet and Hecla, which last has done more to demonstrate the wealth of the copper district than all of the other mines combined.

At this stage, and including the results following during two or three years, the real substantial interest in copper mining commences. The early history of enterprise in that direction has only been sketched, in order to lead up to the practical questions involved in the present consideration of the subject.

Having already admitted that great extravagance attended early mining, it is proper to state that now there is no such objection to be urged. Experience has taught its dearly-bought lessons, and the managers have learned to direct the expenditure of money economically and judiciously. Invention also has come to the aid of mining, and now the slow and laborious process of hand-drilling has been superseded by the use of steam and compressed air. Improved stamps have been added, and various devices for saving the copper and for the more ecomonical handling of the mineral have been introduced, while the means of transporting supplies and of copper have been greatly increased. But perhaps the most extraordinary result of all is in the character of the product and its treatment, which have satisfactorily demonstrated that mines which yield but little mass copper, but are rich in stamp rock, are the most productive and profitable. During the years when the Cliff and Minesota were successfully wrought, a mine which produced only stamp stuff, yielding even 4 or 5 per cent. of copper to the ton of rock, would not have been considered worth working, while now the greatest product by far is obtained from mines producing this stamp mineral almost exclusively.

The mines which are now paying dividends are as follows:

The Calumet and Hecla,	which produced in		1872,	9,718	tons.
" Quincy,	"	"	"	1,402	"
" Franklin and Pewabic,	"	"	"	630	"
" Copper Falls,	"	"	"	423	"
" Central,	"	"	"	803	"
" Minesota,	"	"	"	150	"
" National,	"	"	"	251	"
" Ridge,	"	"	"	170	"
				13,547	tons.

Of these the Calumet, Quincy, Franklin, Pewabic and Copper Falls, producing together 12,173 tons of mineral out of 13,547 tons, derived their profits almost entirely from stamp copper, which, upon the richest (the Calumet and Hecla) does not average over five per cent of copper to the ton of rock. This is a most extraordinary revolution in copper mining; and it appears still more wonderful when it is added that the Quincy stamp rock produces but an average of 2 14-100 per cent., while the Copper Falls claims but an average of about one and one-half per cent. The fact is thus established, that mines economically and judiciously worked, and producing but one and one-half per cent. of copper to the ton of rock, or thirty pounds of metal to two thousand pounds of stamp stuff, can pay good returns upon money invested, with copper ruling in market value much below present prices. This fact stands out the most prominently of any in favor of the great value of the Lake Superior Copper Mines, and is by far the most encouraging feature in the history of that district,—demonstrating as it does the practicability of working profitably hundreds of mines which have been partially abandoned, and encouraging attempts to bring many of them to dividend payments by patient development and well-considered expenditures. Not much stress has hitherto been laid upon this striking fact, but it will impress itself strongly upon all who have followed the history of copper mining to its present development.

The next feature, bearing upon copper mining in its present aspect, is the market price of copper and its probable consumption. The market value of this metal must of course fluctuate, but the minimum of cost can be easily established by the working of former years. Its present cost at the mines varies according to the per centage of ingot copper to the mineral rock; thus, while the Calumet and Hecla may make a good profit with copper at eighteen cents per pound, it would seriously interfere with the profits of some, and oblige others to cease operations entirely. This is an argument which applies to all industrial enterprises—profits always being regulated by supply and demand to a large extent. At present, however, the demand for copper throughout the world, its increased and increasing use for various purposes, and the small gain in the product, point to a price sustained to paying points for some years. In this connection it is worthy of note that Lake Superior copper, from its better quality and adaptability for special purposes, commands a price from three to four cents per pound higher than that imported. It is a well-known fact, also, that the consumption of this metal has been stimulated to a very large extent by various manufactures, such as the metallic cartridge; and while the mining men of this country have no desire to see any further material advance in the market price of ingot copper, they have no great reason to fear any important decline. Taking into account the mutual advantage of producers and consumers, it would probably be more advantageous to both that prices should not advance enough to check its use, however much present profits might be increased. This is the view taken by most of the intelligent men interested in copper mining.

Having thus stated briefly the salient points of this great interest, it is not out of place to note their bearing upon the small and comparatively unfledged concerns which have so long lingered in doubt, and which are occasionally brought out in the stock market as competitors for public favor.

It is somewhat unpopular to express any ideas in favor of these much abused concerns, called, in street parlance, "low priced coppers," and much of the criticism bestowed upon them in former times has doubtless been just; but the facts developed within a few years, and to which brief allusion has been made, cannot be pushed aside by sneers which are based upon ignorance. Success in mining is not to be attained in a day or a month, nor are the schemes of adventurers, which always follow in the wake of enterprise, to be received as full condemnation of the efforts of those who believe in the prosecution of legitimate mining. The pioneers at Lake Superior were at first assured by eminent geologists that the existence of copper in a native form was impossible—but they persisted, notwithstanding, and developed the Cliff and Minesota as facts which were stronger than geological theories. So the mines which are turning out millions from stamp work, controvert the positive opinions of some years since, that stamp rock yielding but a per centage of from 1 to 5 per cent. could not possibly pay. The curse of legitimate mining is the greed and scheming of speculators, who impose false hopes upon the public and inculcate the idea that investments will bring immediate returns, and thus stimulate speculations which have no sound basis.

Fortunately, the sentiment of the times is against this indiscriminate speculation upon public credulity, and the first inquiries now in regard to a copper mining stock are as to its location, veins and prospects of working. The popular idea is, that if a mine is not worth prosecuting, it is not worth buying, unless for mere farming or timber land. Consequently, the "low priced coppers" most in public favor are those which are showing a disposition to develope whatever value they may have.

The facts given in the preceeding pages demonstrate the practicability of bringing into life and profit many of the "low priced coppers," but it should be done carefully, intelligently and earnestly—having in view substantial results. If such ideas can be inculcated and carried out, many of these low priced concerns, which are now made footballs in the stock market, can be started upon a sound and respectable basis, with a promise of good results. The success of the Calumet,

Quincy, Copper Falls, Pewabic, Franklin, Central, and Ridge fully demonstrate this.

The first step which should be taken by companies possessing, as the managers have reason to believe, good mining locations, is that of reorganization, in which process men of character and standing should replace boards of directors which have been elected merely to preserve organization, and who take but little interest in the management. The second, to secure men to fill the offices of President and Treasurer who will devote time and intelligence to the affairs of the company.

These two indispensable measures having been carried, the directors should authorize thorough surveys of the property and the employment of competent and trustworthy men to advise as to the prospects of profitable mining. The expenses of all this preliminary work would be small, and if money is lacking, a trifling assessment would defray the expense.

The surveys and investigations having been completed, the directors can then decide whether the prospects warrant a further expenditure to prove productiveness. *Expensive surface improvements* should not be incurred until the mining prospects are considered excellent,—one of the great errors in former mining having been to spend large sums in preparation for a business which was merely in anticipation.

Many of the low priced mines have extensive surface improvements and costly machinery already provided, and if investigation proves them to be worth working, the expense of starting would be small. Among such mines are the Star, Toltec, Pontiac, &c. Upon these mines a large amount of money has been expended, and with such an important expense already incurred, it would require but little money to prove them. Others which have no "plant"—to use a mining term—should confine their expenditures at first to the proving up of veins.

The Star is a fair example of the first named. This mine is well equipped and in good condition to prepare copper for market, when it is found in paying quantity, and the process recommended is precisely what the company is pursuing.

The Petherick is a fair instance of the latter class, having probably the same copper deposits as the Copper Fall's. The managers propose to prove them, before expending money in surface improvements. When the development warrants it, the company can command more money to carry out the work.

The Allouez is a fair illustration of the public appreciation of earnest work upon a location which had for years the reputation of owning very promising veins, but which until within the last eighteen months were suffered to remain unworked. Believing in good results, the managers have boldly levied assessments amounting to $200,000 during the time and are vigorously pushing developments with excellent prospects of success, and the market value of the stock at the present time, with the assessments of $10 per share added, is much higher than before.

To encourage energetic efforts and a liberal expenditure of capital in prosecuting copper mining on Lake Superior, we have the astonishing success of the Calumet and Hecla, the richest mine, as established by the value of its annual product, of any in the world—whether of gold, silver or copper.

In 1872 it produced 9,718 tons of mineral, or at 90 per cent. 8,747 tons, or 17,494,000 lbs. of ingot copper.

At an average of 30 cents per pound this would yield - - - - - - - $5,248,200

Calculating its cost at 12 cents per pound, the expense of this would be - - - $2,099,280

Or a net profit of - - - - - - $3,148,920

This calculation is based upon unofficial statements, but that it cannot be far from the actual results is proved by the fact that the mine paid $2,800,000 in cash dividends to its stockholders during the year,

Here is a single mine producing more than two-thirds of all the copper mines in the United States, and paying a larger profit than any single gold, silver, or other mine in the world, and nearly all from stamp rock, yielding an average of not more than five per cent., or 100 *pounds to the ton.*

And this astounding result has been obtained from a location which, but a few years since, was considered quite as doubtful as a large number of the low-priced copper stocks dealt in at the Stock Exchange. Such a product in any gold or silver district in the world would create wild excitement, and millions of capital would be attracted to the locality; but the early history of copper mining, which has been briefly given in the preceding pages, has prevented this solid argument from having its effect,—and the sluggish movements in other enterprises in the copper district have naturally contributed to blind people to the advantages and profits of copper mining.

To expect a Calumet and Hecla from many of the other locations upon that mineral range would be, of course, absurd; but to assume that there may not be others quite as rich and productive as the Calumet and Quincy, would be quite as much so.

The great, all-important facts are, that we have these rich developments, and that the profits are obtained *almost entirely from stamp rock*, which, a few years since, would have been rejected unless accompanied by masses and barrel work. When the Cliff mine was at the height of its prosperity, it is stated upon good authority, that the orders of the mining agent were to throw away stamp rock not yielding over three per cent.

There is encouragement enough in these great successes to put life into every copper mine in Lake Superior which can be proved by exploration to be worthy of opening, and in time this will be more forcibly impressed upon capitalists by further developments; but meanwhile it cannot be expected that money will seek investment in so-called mines, which are allowed to rest undisturbed by the tools of the miner, and which have not even had the advantage of preliminary exploration. The price of copper favors mining enterprise, and the great facts established in regard to the profits on stamp rock are sufficient for every company to begin work in earnest, at least so far as to demonstrate that it is not enjoying a fictitious life, in floating upon the current with meritorious and profitable concerns, or those which are expending money in opening promising veins. Under all the circumstances, if the managers of co-called "copper mines" do not soon show a disposition to take the first steps toward development, which will involve but a small expense, it will naturally be concluded that they have no real confidence in the value of their locations for copper mining.

People who choose to deal in copper stocks, and desire to place $200, $300 or $500 in the venture of a hundred shares, should make careful inquiries as to the location, veins and prospects of actual mining, and must finally prefer to take the chances in those which it is seriously proposed to open. If these conditions do not exist, the investment will be purely speculative, resting simply upon the credit of other mines, of which they are but the reflection.

ASSESSMENTS AND DIVIDENDS.

ASSESSMENTS IN 1872.

Allouez	$160,000
Petherick	20,000
St. Clair	10,000
Total	$190,000

[From the Portage Lake Mining Gazette.]

TABLE OF ASSESSMENTS LEVIED,

AS NEARLY AS CAN BE ASCERTAINED, SINCE THE COMMENCEMENT OF OPERATIONS, IN 1845.

Adams	$100,000	Copper Creek	$30,000
Adventure	100,000	Copper Falls	510,000
Ætna	140,000	Copper Harbor	20,000
Albany & Boston	156,000	Dana	68,000
Algomah	65,000	Dacotah	56,500
Allouez	298,000	Delaware	350,000
American	20,000	Devon	20,000
Amygdaloid	470,000	Dorchester	30,000
Arnold	30,000	Douglass	130,000
Atlas	40,000	Dover	20,000
Aztec	150,000	Dudley	20,000
Bay State	385,000	Eagle Harbor	80,000
Bohemian	343,000	Eagle River	85,000
Boston	45,000	Edwards	32,500
Caledonia	140,000	Empire	76,000
Calumet	300,000	Evergreen Bluff	150,000
Central	100,000	Everett	20,000
Concord	120,000	Flint Steel River	244,000

Franklin	$370,000
Frue	50,000
Garden City	156,000
Girard	43,300
Grand Portage	50,000
Great Western	40,000
Hamilton	40,000
Hancock	530,000
Hanover	33,000
Hartford	30,000
Hecla	500,000
Highland	20,000
Hilton	50,000
Hope	22,000
Hulbert	15,000
Humboldt	100,000
Hungarian	20,000
Huron	500,000
Indiana	200,000
Iroquois	20,000
Isle Royale	910,000
Kearsarge	40,000
Keweenaw	100,000
Knowlton	160,000
Lake Superior	40,000
Madison	120,000
Mandan	65,300
Manhattan	110,000
Mass	98,800
Medora	38,400
Mendota	147,500
Merrimac	117,900
Mesnard	159,000
Michigan	40,000
Milton	30,000
Minesota	436,000
National	110,000
Native	39,000
Naumkeag	20,000
North Cliff	$110,000
Northwestern	227,300
Norwich	230,000
Ogima	140,000
Ossipee	90,000
Pennsylvania	500,000
Petherick	70,000
Pewabic	235,000
Philadelphia & Boston	24,300
Phœnix	820,000
Pittsburgh & Boston	110,000
Pontiac	104,900
Quincy	200,000
Reliance	20,000
Resolute	51,000
Ridge	200,000
Rockland	340,000
Rhode Island	100,000
St. Clair	140,000
St. Louis	20,000
St. Mary's	110,000
Salem	10,000
Seneca	40,000
Sharon	2,000
Sheldon & Columbian	460,000
South Pewabic '69	500,000
South Side	90,000
Star	265,000
Superior	220,000
Toltec	420,000
Tremont	22,000
Victoria	85,000
Vulcan	30,000
Washington	20,000
West Minesota	45,000
Winona	20,000
Winthrop	90,000
Schoolcraft	380,000
Total	$17,296,500

Dividends in 1872.

Calumet and Hecla	$2,750,000
Quincy	350,000
Pittsburgh & Boston (Cliff)	100,000
Central	80,000
Minesota	50,000
Franklin	20,000
Pewabic	20,000
National	20,000
Total dividends	$3,390,000
Total assessments	190,000
Excess of dividends over assessments	$3,200,000

Total Dividends Declared.

Calumet and Hecla	$4,800,000
Pittsburgh and Boston (Cliff)	2,280,000
Minnesota	1,750,000
Quincy	1,490,000
Central	550,000
Pewabic	400,000
National	300,000
Franklin	240,000
Copper Falls	100,000
	$11,910,000

[From the Portage Lake Mining Gazette.]

PRODUCTS OF THE COPPER MINES IN 1872.

Portage Lake District.

	Tons.	lbs.
Calumet and Hecla	9,717	1,713
Quincy	1,402	954
Franklin and Pewabic	630	
Houghton	300	
Schoolcraft	300	
Isle Royale	81	
Concord	80	
Hancock	26	
Other sources	6	
Total	12,543	667
Product in 1871	12,791	1,595
Decrease in 1872	248	928

Keweenaw Point District.

	Tons.	lbs.
Central	803	1,402
Phœnix	477	
Copper Falls	423	987
Delaware	110	161
Cliff	74	1,087
St. Clair	13	353
Petherick	6	146
Other sources	8	
Total	1,916	136
Product in 1871	2,340	1,391
Decrease in 1872	424	1,255

Ontonagon District.

	Tons.	lbs.
National	251	1,145
Ridge	170	
Minesota	159	1,284
Flint Steel River	37	422
Bohemian	36	1,179
Rockland	25	1,216
Knowlton	11	1,300
Adventure	8	50
Aztec	2	303
Victoria	2	1,904
Mass	1	572
Total	706	1,366
Product in 1871	938	1,407
Decrease in 1872	232	41

Recapitulation.

	Tons.	lbs.
Houghton County	12,543	667
Keweenaw County	1,916	136
Ontonagon County	706	1,366
Total	15,166	169
Total in 1871	16,071	393
Decrease in 1872	905	224

Statemeft of Copper (Mineral) Product from 1845 to 1873.

	Tons.
1845 to 1854	7,642
1854 to 1858	11,312
1858	4,100
1859	4,200
1860	6,000
1861	7,500
1862	9,962
1863	8,548
1864	8,472
1865	10,791
1866	10,376
1867	11,735
1868	13,049
1869	15,288
1870	16,183
1871	16,071
1872	15,166
Total	176,395

Approximate Statement of Ingot Copper Produced, and its Valu.

	Tons.	Value.
1845 to 1858	13,955	$9,000,500
1858	3,500	1,886,000
1859	3,500	1,890,000
1860	4,800	2,610,000
1861	6,000	3,337,500
1862	8,000	3,402,000
1863	6,500	4,420,000
1864	6,500	6,110,000
1865	7,000	5,145,000
1866	7,000	4,760,000
1867	8,200	4,140,000
1868	9,935	4,592,000
1869	12,200	5,368,000
1870	12,946	5,696,240
1871	12,857	6,171,360
1872	12,132	7,774,720
Total	135,075	$76,303,320

MARKET FOR COPPER.

The following information in reference to the market for copper and stocks on hand, is derived from the excellent "Metal Circulars" of Messrs. White & Haskell, of New York, and Van Dadelszen and North of London.

NEW YORK, December 31, 1872.

COPPER.—The year closes with small stock of all kinds and market strong and active, at 33 1-2c. bid and 34c. asked for Lake on spot, and 34c. bid for January, February and March, 34 1-2c. asked. English, Best Selected, 29c. bid, 29 1-4c. asked. Baltimore, none offering for immediate delivery, 33c. asked for January, February and March.

The stock of Ingot, January 1, 1872, estimated .	5,000,000 lbs.
The estimated production of 1872, Lake Superior,	23,500,000 "
" " " " Tennessee, .	1,500,000 "
" " " " Other American Smelters, ...	3,000,000 "
The estimated Imports in 1872, of English B. S. and Tough Ingot, ...	6,000,000 "
	39,000,000 "
The estimated delivery for consumption, 1872. .	34,000,000 "
Stock January 1, 1873, ...	5,000,000 "

In consequence of small stocks and a strong foreign market, the year 1872 opened firm at 27 1-4c., for Lake, advancing during January and February to 28 1-2c. with sales of about 3,000,000 lbs. in January, and about 2,000,000 lbs. in February.

LONDON, January 6, 1872.

COPPER.—A large business has been done, and prices advanced during the month fully £6 per ton. The anticipation expressed in our last has thus been fully realized, and from the nature of the buying we look with confidence to the future.

	Dec. 1,'72	Jan. 1, '73	Jan. 1, '72	Jan. 1, '71
	Tons.	Tons.	Tons.	Tons.
Stock, Liverpool and Swansea	23,400	23,400	11,300	24,000
" Havre	180	286	4,015	7,678
" London	7,300	7,700	3,464	4,035
	30,880	31,386	18,779	35,713
Chili Produce, afloat and chartered....	9,400	9,900	12,000	11,600
Total.............	40,280	41,286	30,779	47,313

The stock of copper at railway stations in Birmingham is now estimated at only 500 tons; in years past, it used to be upwards of 2,000 tons. The quantity of Australian now afloat is estimated at only 750 tons, against in former years above 2,500 tons. The stocks in Chili, inclusive of one month's production, are now estimated at 3,750 tons, against 10,750 tons last year.

Statement of the Locations, Condition, &c., of

Lake Superior Copper Mines.

Mainly Derived from Official Sources.

MINES NOW PAYING DIVIDENDS.

CALUMET AND HECLA MINING COMPANY.—80,000 SHARES.

Portage Lake District. Product in 1872, 9,718 tons. Profits of 1872 estimated at about $3,000,000. Dividends paid within fourteen months, $2,800,000.

These figures are not official, but are considered tolerably accurate. The policy of the managers seems opposed to public statements of the company's business. It is stated by men well informed in regard to this mine, that it is opened sufficiently to give assurance of quite as large a product as that of 1872, for three years to come.

QUINCY MINING COMPANY.—(Official.)—20,000 SHARES.

Portage Lake District. Mining location 626 acres, near the Pewabic, Franklin and Isle Royale mines. Working on the Pewabic and Quincy veins. The Pewabic lode has been opened below the 210 fathom level—(1260 feet); the Quincy to the 30 fathom level (180 feet.) The amount of copper taken out is 28,716,297 pounds. The product of 1872, 1402 1-2 tons. Profits of 1871 and 1872 about $400,000. Dividends of 1871 and 1872, $490,000—embracing a part of the surplus of previous years. Surplus in the treasury after payment of the dividend of February 24, about $250,000. Extensive surface improvements and machinery; mine fully equipped. Amount assessed on stockholders prior to dividends, $200,000.

PITTSBUGH AND BOSTON (CLIFF) MINING CO.—20,000 SHARES.

Keweenaw Point District. Famous in the early history of Lake Superior Copper Mines. The first mine which paid a dividend. Mining work was suspended several years, but within eighteen months the mine has been sold to M. H. Simpson, Esq., of Boston, for $100,000, and it is now worked by him. Extensive surface improvements and ample machinery. Total amount of assessments levied, $110,000. Total amount of dividends $2,280,000. Stock sold in the market in its prosperous years at $160.

CENTRAL MINING COMPANY.—(Unofficial.)—20,000 SHARES.

Location, Keweenaw Point District. Product 1872, 804 tons. Fully equipped. Extensive surface improvements and machinery. Dividends in 1872, $80,000. Total amount of dividends, $550,000. Total amount assessed on stockholders, $100,000.

MINESOTA MINING COMPANY.—20,000 SHARES.

Ontonagon District. The second mine to pay dividends on Lake Superior. Very prosperous in its early history. Work suspended for several years, but now worked on tribute. Total dividends, $1,750,000. Dividend of 1872, $50,000. Extensive surface improvements and machinery. Total amount assessed on stockholders, $436,000.

COPPER FALLS MINING COMPANY.—(Official.)—20,000 SHARES.

Location, near Eagle Harbor. About 2,000 acres; near the Central and Petherick mines. Working the Ash-Bed and Owl Creek veins. Mines extensively opened, and fully provided with all necessary surface improvements and machinery, including four heads of Balls Stamps. Product in 1872, 423 1-2 tons. Dividends February 10, and August 21, 1871, one dollar per share. Dividends suspended in 1872, in consequence of the falling in of a portion of the mine, which absorbed the surplus on hand. Recent working upon the Owl Creek vein has developed very rich deposits of copper, and the product will, it is believed, be largely increased in 1873.

RIDGE MINING COMPANY.—(Official.)—20,000 SHARES.

Ontonagon District. 1,393 acres. Product of 1872, 170 tons. Profits of 1871 and 1872, $75,000. Dividend of 1872, $50,000. Surplus, after payment of dividend, in copper and cash, $70,000. Mine well supplied with all necessary machinery, and a stamp mill is in course of erection.

PEWABIC CONSOLIDATED MINING CO., (Official,) 20,000 SHARES.

Portage Lake District. About 1,500 acres. Near the Franklin and Quincy mines. Veins consisting of the Pewabic, Amygdaloid, A and B Conglomerate. Mine worked extensively. Total amount of Ingot Copper taken out, 18,928,079 pounds, or 9,464 tons. Extensive surface improvements and ample machinery. No debt. Assets, $33,-536.61,—consisting of cash and receivables, $18,268.30. Copper and supplies, $14,136.83. Profits of 1871 and 1872, $65,627.19. Dividends of 1871 and 1872, $60,000.

The mines of this company, consisting of the Pewabic and Concord, are worked on tribute. The total product of the Pewabic in 1871, was 522,861 pounds ingot copper, of which the company's share, under the tribute contract, was 74,936 pounds ; that of the Concord, 158,000 pounds, of which the company's share was 15,800 pounds, making a total of 90,736 pounds as the company's proportion. Assessments $185,000.

FRANKLIN MINING COMPANY.—(Official.)—20,000 SHARES.

Portage Lake District. About 180 acres, near Pewabic and Quincy Mines. Veins old Pewabic, and Boston and Albany Conglomerate. Mine extensively opened and well equipped. Amount of copper taken out, 16,421,853 pounds, or 8,211 tons ingot. Debt, $9,278.72. Assets cash and loans, $8,876.56. Copper and supplies, $9,917.44. Total, $18,794. Profits of 1871 and 1872, $51,795.79. Dividends of 1871 and 1872, $60,-000. Worked on tribute since July 1, 1872.

NATIONAL MINING COMPANY.—(Unofficial.)—20,000 SHARES.

Ontonagon District. Mine well opened. Worked on tribute system. Divident in 1872, $20,000. Total dividends, $300,000. Total amount of assessments, $110,000.

Mines Not Now Paying Dividends.

Statements Mainly Derived from Official Sources.

ÆTNA MINING CO.–20,000 SHARES.

Location, Keweenaw Point District. Owned in Philadelphia. About to commence work.

ALBANY AND BOSTON MINING CO.—20,000 SHARES—(Official).

Location, Portage Lake District. 1760 acres. Midway between the Quincy and Çalumet and Heçla Mines. No debt. A small amount of money in the treasury. The company owns a valuable mill site and water front on Portage Lake. Not at present working.

ALGOMAH MINING CO.—20,000 SHARES.

Ontonagan District. Adjoining the Toltec Mine. Small amount of work done.

ALLOUEZ MINING CO.—20,000 SHARES—(Official.)

Location, Keweenaw Point District. Two miles north of the Calumet and Hecla. 2,738 acres. Veins,—Allouez, Conglomerate and Calumet and Hecla Conglomerate. Large amount of work done. 1150 feet of drifting and 740 feet shafting. Cash assets about $40,000. This mine is held in great esteem by those interested, and is considered to be one of the most promising of those in process of development.

AMYGDALOID MINING CO.—20,000 SHARES.

Location, Keweenaw Point District. Considerable done. Assessments of $470,000.

ATLAS MINING CO.—20,000 SHARES—(Official.)

Location, Keweenaw County. About 360 acres, near the Phœnix and Cliff Mines. Not much work done. A few miners houses built. No debt. About $1,000 in cash in the treasury. Not at present working.

AZTEC MINING CO.—20,000 SHARES.

Location, Ontonagon District. Near Toltec and Bohemian. Some Copper taken out. Fair amount of surface improvements and considerable machinery, engines, stamps, &c. About to reorganize with a view of resuming work.

DANA MINING CO.—20,000 SHARES—(Official).

Location next to Northwestern, which joins the Central. About 900 acres. Was working several years ago upon a very promising vein, by driving an adit under the greenstone. Recently assessed 50 cents per share, which will furnish the treasury with $10,000. Private accounts speak well of the Dana Mine. Not at present working.

DELAWARE COPPER CO.—20,000 SHARES —PAR VALUE, $25.—(Official.)

Location, Keweenaw Point District. 720 acres of mineral and 640 acres of wood land. Several fissure veins exposed, of which the Delaware is now being worked. Letters dated in December report important developments, one mass having been exposed of about 150 tons. Product in 1872, 110 tons. Extensive surface improvements, with stamp mill of 48 heads, capable of stamping and working 150 tons of rock per day. A very promising mine.

EAGLE RIVER MINING CO.—20,000 SHARES.—(Official.)

Location, Keweenaw Point District. Adjoining St. Clair mine.

EVERGREEN BLUFF MINING CO.—20,000 SHARES—(Official).

Location, Ontonagon District. 1960 acres, adjoining the Ridge Mine. Veins same as worked by the Ridge. A large amount of work done. 2,150,000 lbs. of Copper taken out from 1863 to 1869. Extensive surface improvements, consisting of engines, stamp mill, and houses, costing $3,000. No debt. Assets in cash, $7,000. Copper and supplies estimated at $10,000. The company has 2,400 feet of the vein now worked by the Ridge. Not worked since 1869.

FLINT STEEL RIVER COPPER CO.—40,000 SHARES, $15 PER SH.—(Official.)

Location, Ontonagon District. 2034 acres. Between Minesota and Ridge Mines. Veins,—Flint Steel, Champion, Knowlton and Ridge. Mine opened to the extent of 5,000 feet. 983,440 lbs. of ingot copper taken out. Mine well equipped with stamp mill, engines, buildings, &c., valued at $80,000. Debt, $6,000. Assets, $17,000; consisting of cash $2000, Copper and supplies $15,000. Working with good prospects.

FLINT STEEL RIVER MINING COMPANY,—20,000 SHARES.

Location Ontonagon District. Adjoining the Rockland and Superior. Considerable work done. $244,000 levied in assessments. Considered very promising when the Minesota had so high a reputation.

GLOBE COPPER CO.—20,000 SHARES.—(Official.)

3,200 acres, located south of the property of the Atlantic Mining Company, in Houghton County, Michigan. Carries the same class of veins as exposed on the Atlantic. No work done except surface explorations. No machinery. No surface improvements. No debt. Assets equivalent to about $3200 cash. The Globe property is five miles in length by one mile in width, and covers fully two-thirds of the entire width of the metalliferous formation.

HANCOCK MINING COMPANY,—20,000 SHARES.

Portage Lake District. Adjoining Quincy. Extensively opened, and large surface improvements. Mine sold and now worked by individuals, who are about to organize a new company. The mining of the past year said to have been quite successful.

HANOVER MINING COMPANY,—20,000 SHARES.

Location, Keweenaw Point District, in the vicinity of the Clark mine, about three miles from Copper Harbor. Very little work done. No surface improvements. $33,000 levied in assessments. Said to have a very promising vien. The Company has been reorganized, and an exploration is to be made as soon as the snow leaves the ground.

HIGHLAND COPPER CO.—20,000 SHARES.—(Official.)

Location, Portage Lake. About 1,000 acres. Near Franklin and Pewabic mines. The Company has several veins, upon which but little work has been done, and some surface improvements in the way of buildings have been made. No debt. About $1,000 cash in the treasury. Not at present working.

HUMBOLDT MINING CO.—20,000 SHARES.—(Official.)

Location, near the Petherick and Copper Falls. Same line of Ash-Bed and veins. Vein about 3 feet wide, opened 40 to 50 feet, by adit level 8 feet square. While opening on the bluff had a very good show of Copper. Driving adit towards the Ash-Bed about the same as the Petherick. About $6,000 in the treasury. Expending from $500 to $600 per month. A mine of good promise.

ISLE ROYALE MINING CO.—20,000 SHARES.—(Unofficial).

Location, Portage Lake District. Mine extensively opened and worked, and a large amount of money ($910,000) expended. No returns received from the Company. In the early history of the Portage Lake Mines the Isle Royale was one of the most conspicuous, and a favorite in the stock market. Product in 1872, 81 tons.

KNOWLTON MINING CO.—20,000 SHARES.—(Official.)

Location, Ontonagan District. 680 acres. Near the Evergreen, Bluff, and Ridge Mines. The Knowlton vein has been opened down to the fourth level. The company has a stamp mill, engines and houses. 1,086,000 lbs. ingot copper taken out from 1863 to 1869. No debt. Cash in the treasury, $500. Other assets in Copper and supplies, $3000, and 340 shares of the Company's stock. Mine now worked on tribute, the Company receiving one quarter of the whole product.

MADISON MINING CO.—20,000 SHARES.—(Official.)

Location, Keweenaw County. About 2,000 acres near the Central and Copper Falls Mines. No debt. About $500 in cash in the treasury. Not at present working.

MANHATTAN MINING CO.—20,000 SHARES.

Location, about 4 miles south of the Cliff mine, bordering on the same mineral range. 440 acres. Has a number of buildings in fair condition. Considerable work done, with fair profits. Not at present working.

MESNARD MINING CO.—20,000 SHARES.

Portage Lake District. Adjoining Franklin and Pontiac. Pewabic and Franklin lodes. Considerable work done. Fair surface improvements. About $10,000 in the treasury. Recently leased on the tribute system.

PETHERICK MINING CO.—20,000 SHARES.—(Official).

Location, Keweenaw County. About 1100 acres. Formerly the Copper Falls Mine. Has the Ash-Bed deposit of the Copper Falls and the so-called Petherick vein. Good surface improvement. Working on vein from base of bluff, intending to strike the Ash-Bed of the Copper Falls at 2,000 feet distance. This vein is the same worked by John Uren, from which he took considerable mass copper. Adit now driven about 50 feet. Further contracts at $10 per foot. When the adit reaches the Ash-Bed it will drain the mine and thus save hoisting and pumping expenses. Cash on hand, about $13,000. Expending about $1,500 per month. Considered promising by the managers.

PHŒNIX COPPER COMPANY.—20,000 SHARES.—(Official).

Location, Eagle River, Keweenaw County. About 1900 acres. Near the Cliff, Atlas and Boston, Humboldt, &c. Veins opened,—Old Phœnix, New Phœnix and Robbins, also the belt known as the Ash-Bed. Operations at present confined to the New Phœnix, south of Greenstone, in ground similar to that worked by the Cliff. Present working shaft down about 700 feet, between the 80 and 90 fathom level. Vein showing well in the belt which yielded the large mass in 1869. The Robbins vein, in the same vein of ground, is very promising. The company has over a mile in length of the Ash-Bed, and belts corresponding with those which have recently shown so well in the "Owl Creek" vein at Copper Falls. Liabilities Jan. 1st, $66,836.80. Assets in cash, Copper and supplies, $131,839.05. Valuation of surface improvements, $135,112.42. Profit in 1871, $92,252.44. Profit in 1872, $28,327.64.

In 1871 great exertions were made to produce a large amount of Copper, and the mine not left in a satisfactory working condition—compelling work in opening in 1872 and the expenditure of $35,000 in surface improvements. The present condition of the mine is very satisfactory, and it is showing well for mass copper and stamp rock. The Phœnix is under conservative and trustworthy management, and is considered one of the most promising on the list.

PONTIAC MINING CO.—20,000 SHARES.

Portage Lake District. Adjoining the Mesnard. Considerable work done, and fair surface improvements. About $3000 in the treasury. The managers expect to resume work soon.

ROCKLAND MINING CO.—20,000 SHARES.—(Unofficial.)

Ontonagon District. Total amount of assessments, $360,000. Near Minesota and National. Extensively opened. Now working on tribute system.

SCHOOLCRAFT MINING CO.—20,000 SHARES.—(Unofficial.)

The Schoolcraft is working on what is believed to be the great vein of the Calumet and Hecla. In 1872 it produced 300 tons.

SENECA MINING CO.—20,000 SHARES.—(Official.)

2,740 acres, located in Keweenaw County, Michigan, near the Allouez mine. It carries all the discovered productive belts of Conglomerate and Amygdaloid. There is also a fissure vein exposed, which for surface indications promises equal to any vein of the kind ever opened in the country. But little work has been done up to this time, but it is probable that active operations will be commenced next summer. Surface improvements consist of four substantial log buildings for dwellings. No debt. Cash assets about $5,000.

SOUTH SIDE MINING CO.—20,000 SHARES.—(Official.)

Location, Portage Lake. 175 acres. Several good buildings, dwelling houses, shops, &c. Considerable work done in driving adits without important results. Near the Naumkeag and Dacotah. Amount assessed, $70,000. Not at present working.

STAR MINING CO.—20,000 SHARES.—(Official.)

Location near Copper Harbor, adjoining the Clark Mine, now working with excellent prospects under the direction of Mr. Estevant. Excellent water power, 16 heads of stamps, and all needed surface buildings. The old vein worked ten years since was showing well. Now driving adit in the Clark vein. Recent letters speak encouragingly of propects. About $5,000 in the treasury. Expending about $600 per month.

ST. CLAIR MINING CO.—20,000 SHARES.—(Official.)

Location, Keweenaw County. 160 acres. Near Phœnix and Bay State. One transverse vein, worked to a depth of 300 feet and well opened. Amount of copper taken out produced $86,000. Very good equipment. No debt.

ST. LOUIS COPPER CO.—20,000 SHARES.—(Official.)

Portage Lake District. Near Calumet and Hecla. About 870 acres. Vein called the "St. Louis lode." Some buildings have been erected. No debt. Assets, $2,500 in cash. Not at present working. The "St. Louis lode" is considered very promising, and according to good authority shows heavy copper in many places.

St. Mary's Mining Co.—20,000 Shares.—(Official.)

Location, Portage Lake District. About 800 acres. Near Quincy, Pewabic, and Franklin mines. Veins,—Pewabic, Epidote and Conglomerate. Not much work done. A few tons of copper taken out. Surface improvements. Few houses. No machinery. No debt. About $1,000 cash in the treasury. Not at present working.

Superior Mining Co.—20,000 Shares.

Ontonagon District. Near Rockland and Flint Steel River Mines. Considerable work done.

Toltec (Consolidated) Mining Co.—20,000 Shares.

Ontonagon District. 1120 acres. Near Bohemian and Aztec Mines. Extensively opened and at one time considered very promising. Large surface improvements and machinery, including stamps, engine, &c. $420,000 levied in assessments. Has other veins. Reorganization contemplated, and the resumption of work under consideration.

Washington Copper Co.—20,000 Shares—(Official.)

Location, Keweenaw Co. About 1000 acres near the Central and Copper Falls Mines. Not much work done. Debt about $2,000. Not at present working.

Winthrop Mining Co.—20,000 Shares.—(Official).

Location, adjoining the Central Mine. About 640 acres. Surface improvements of buildings, engines, &c. Some work done with a fair show. No money in the treasury. Not at present working.

NOTE.—The compiler of these mining statistics would gladly have added authentic accounts of many other mines worthy of mention, but it has been impossible to procure them in time for this edition.

If it is concluded to publish another edition, any statements of mining companies on the Michigan peninsula will be added, if the managers will take the trouble to forward them to the undersigned.

O. D. Ashley,

P. O. Box 560. No. 4 Congress Square, Boston.

Certificates.

BOSTON, March 20th, 1873.

O. D. ASHLEY, ESQ., 4 Congress Square:

DEAR SIR,—I have received and perused with much care your pamphlet descriptive of "The Copper Mines of Lake Superior." The information you give is not only valuable but entirely trustworthy, and cannot fail to excite a degree of interest in the minds of capitalists which the subject you treat of so justly merits. Accept my individual thanks for the effort to bring this important matter so ably before the public, and believe me

Very truly and respectfully yours, JOS. W. CLARK.

---

19 BROAD STREET, NEW YORK, March 19th, 1873.

O. D. ASHLEY, ESQ., 4 Congress Square, Boston:

DEAR SIR,—I have received and read your pamphlet giving a description of the early development of the copper resources of the country bordering on Lake Superior, from the first commencement of mining in that country down to the close of the year 1872. I have found its perusal both agreeable and instructive. Its publication is timely, and cannot but have a good influence in placing that great national interest in its true light with those who take the time to read it.

Yours truly, C. C. DOUGLAS.

BOSTON, March 20th, 1873.

O. D. ASHLEY, ESQ., 4 Congress Square:

MY DEAR SIR,—I have read with interest your pamphlet on the Copper Mines of Lake Superior. Having been connected with the Mining enterprise on Lake Superior almost from its inception, I can from my own knowledge attest the accuracy of your statements.

Very truly yours, HORATIO BIGELOW.

BOSTON, March 19th, 1873.

O. D. ASHLEY, ESQ., 4 Congress Square:

DEAR SIR,—Your pamphlet on the Copper Mines of Lake Superior contains much important and valuable information for persons interested in those enterprises, and, as intended, it will give reliable facts as to the present condition of those properties.

Yours truly, H. W. NELSON.

BOSTON, March 17th, 1873.

O. D. ASHLEY, ESQ., 4 Congress Square:

DEAR SIR,—We thank you for your excellent pamphlet on "The Copper Mines of Lake Superior." Its compilation of statistics is correct, and is evidence of great care and labor or your part. Our experience as Brokers, has taught us that such a book will be very useful.

Yours very truly, C. D. HEAD & T. H. PERKINS,
Bankers, 22 Devonshire St.

NEW YORK, March 17th, 1873.

O. D. ASHLEY, ESQ., 4 Congress Square, Boston:

DEAR SIR,—Your interesting pamphlet makes an excellent reference for all those desiring information concerning the Copper Mines of Lake Superior. I believe I have never before seen the statistics so complete and in such a compact form.

Yours truly, R. H. RICKARD.

BOSTON, March 19th, 1873.

O. D. ASHLEY, ESQ., 4 Congress Square:

DEAR SIR,—We have read your pamphlet on the Copper Mines of Lake Superior with interest, and know that it is of great value and gives a vast amount of accurate and reliable information with regard to the mining interests of Lake Superior.

No one can doubt that this country will derive immense benefit from the development of those interests. The untold mineral wealth that exisits throughout the whole Lake Superior region will continue to attract capital and labor for years to come, and an accurate history such as yours is, of what has been done there in the past, cannot fail to be an exceedingly valuable guide as to the course which should be pursued in the future. Yours truly, GEO. W. LONG & Co.

BOSTON, March 18th, 1873.

O. D. ASHLEY, ESQ., 4 Congress Square:

DEAR SIR,—Agreeing with you that there are at Lake Superior many Copper Mines which good management would render profitable, I have read with pleasure your pamphlet, which will, I trust, awaken an interest in legitimate mining in that region of undoubted mineral wealth. Yours truly, J. H. TUTTLE,

Treasurer Phœnix Copper Co.

NEW YORK, March 17th, 1873.

MR. O. D. ASHLEY, 4 Congress Square, Boston:

DEAR SIR.—Your pamphlet on the Copper Mines of Lake Superior has been much needed, and to my mind meets the wants of not only those interested in Copper mining, but of those seeking for investment in Copper stocks. I consider it as very valuable from a statistical point of view, and hope it will be appreciated as it deserves.

Yours very truly, E. H. WHITE,

(White & Haskell, metal brokers, 36 Fulton street.)

SOMERVILLE, March 20th, 1873.

O. D. ASHLEY, ESQ., 4 Congress Square, Boston:

MY DEAR SIR,—I thank you for a copy of your pamphlet relating to the Copper Mines on the peninsula of Michigan. It contains information in condensed form of great value to those interested in operations in that remote and busy region.

Few who have never visited Lake Superior can fully appreciate the amount of mineral wealth, which has in that neighborhood, within a few years, been extracted from the bowels of the earth; promoting, as it has, the establishment and growth of important towns and villages over a country more than a hundred miles in extent; providing a means of livelihood, comfort and happiness to many thousands of the population, and proving a source of revenue to those who have acted with discrimination, economy and skill in a legitimate pursuit of mining. If your labor in this matter results in breaking up the bogus enterprises which cluster around the more promising mines, you will have done the public a great service.

Your ob't servant, JAMES M. SHUTE.

BOSTON, March 21st, 1873.

O. D. ASHLEY, ESQ., 4 Congress Square:

DEAR SIR,—We have read your pamphlet on the "Lake Superior Copper Mines" with much interest, and take pleasure in testifying to its general accuracy. Information of this kind will be very useful in giving to the public the facts about copper mining, and in removing a great deal of the prejudice which exists in regard to it.

Resp'y, &c., yours, BECK BRO'S.

www.ingramcontent.com/pod-product-compliance
Lightning Source LLC
LaVergne TN
LVHW011127110826
845150LV00008B/2270

* 9 7 8 1 4 1 8 1 9 5 5 9 5 *